CONCISE COLLECTION

Planets and Stars

Nicholas Booth

Grange
BOOKS

Published in 1995
by Grange Books
An imprint of Grange Books Plc.
The Grange
Grange Yard
London SE1 3AG

ISBN 1 85627 738 0

Printed in Italy.

Acknowledgments
University of Arizona 41; Astronomical
Society of the Pacific 12; Nick Booth 17, 32,
42, 43; Paul Doherty 46; Jet Propulsion
Laboratory 18, 21, 22; NASA 9, 10, 11, 16, 20,
23, 24; Starland Picture Library 30, 34, 40;
Tass 13.
All artwork supplied by Jack Pelling/Linden
Artists with the exception of Maltings
Partnership 26, 28, 39 and front cover; Paul
Doherty 7 left, 8 right, 14 right, 19 left, 25 left,
27 left, 29 left, 36 right, 38 right. Andrew
Wright 31 left.

Contents

Left: Saturn's rings. Colour variations indicate different chemical composition.

Introduction

Astronomy holds the distinction of being both the oldest and the newest of sciences. 'Oldest' because our earliest ancestors looked to the heavens to try to understand the majestic sights which they believed foretold their destiny. 'Newest' because every month brings startling discoveries, the product of extraordinary techniques and observations at different wavelengths to which the eye is not sensitive.

This book, then, is essentially a photographic album of some of the more unusual objects which populate the universe in which we live. Astronomers sometimes refer to the wide range of objects as the cosmic zoo. Many of these contained here are rare. Our Sun is a pretty humdrum star, yet it is unique because one of its nine planets (ours) contains life.

The order of entries in this book duplicates an imaginary journey outwards from the Sun. The planets and moons of the Solar System are seen first and then the stars, starting with the closer ones, and then whole groupings of stars, known as galaxies.

The most exciting thing about astronomy is that new information redraws those pictures as does the overall interpretation. Today's astro-nomers would like to think that they have the universe figured out, but experience shows this is seldom true. New pieces of evidence can alter our perception of the cosmos rapidly.

Medieval astronomers believed the Earth to be the centre of the universe: the Italian astronomer Galileo made observations with one of the first telescopes that showed otherwise. Renaissance astronomers believed the Sun to be at the centre of the universe: observations with ever more power-ful telescopes showed this to be false. Earlier this century, astronomers realized that the scale of the universe was far larger than they had thought, and that galaxies were zooming away at unimaginable speed. By the start of the Second World War, they thought that there was very little else to add to this picture. The birth of radio astro-nomy and then subsequent astronomy from spacecraft has redrawn that pic-ture once again.

If nothing else, this book will give a guide to the planets and stars as we know them today. Yet a great deal remains unknown. As the disting-uished biologist J.B.S. Haldane once remarked, the universe is not only queerer than we suppose, it *is* queerer than we can suppose.

Note on tabulated information
For some of the entries in this book (particularly the planets) tabulated data is given for the objects shown. In the case of galaxies, figures are not included as they are not well known.

The information in these tables has been simplified as follows:
Average Distance from Sun: The planets do not move in perfectly circular orbits, but elliptical ones. So the average distance is the mean of the farthest distance and closest distance to the Sun. Moons move in elliptical orbits around planets.
Year: The time taken for a planet to complete one orbit around the Sun is technically known as a sidereal year. In the case of the Earth, this is 365.26 days. The figures quoted here are sidereal years, as there are different ways of measuring the way in which an orbit is defined.
Day: The time taken for a planet to make one complete rotation about its axis is known as its day. In the case of the Earth this is 23 hours, 56 minutes and 56.6 seconds.
Mean Temperature: The average temperature on the surface, averaged over the globe. As the Earth has a dense atmosphere, the range in temperature is obviously modified.

The Sun

Period of Rotation: 27.3 days; **Diameter:** 1.392 million km (864,972 miles); **Mass:** 2×10^{27} tonnes; **Surface Gravity:** 28 × Earth

The Sun, our daytime star, is just one of many millions in the Universe and is in no way special. It is a relatively stable, middle-aged star (about 5,000 million years old), and its stability has ensured that life could evolve on Earth and continues to flourish, because ultimately, the Sun is also the source of all our power. At its core, hydrogen is continually being con-

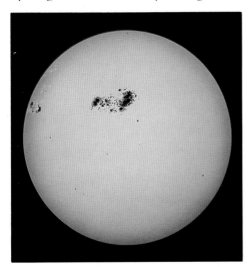

verted to helium by thermonuclear fusion at the rate of 6 million tonnes per second, equivalent to millions of atomic explosions each second.

The internal structure of the Sun is believed to consist of a core whose temperature is around 15 million degrees Celsius (27 million degrees Fahrenheit) surrounded by a region known as the convective zone in which the tremendous heat dissipates upwards. The visible surface of the Sun – the photosphere – is surrounded by a region known as the chromosphere in which violent generation of flares, vast bursts of radiation, occur. The Sun is also generating continuous streams of electrically-charged particles known as the solar wind.

The photosphere is often blemished by dark splotches known as sunspots, which are simply cooler regions. They appear dark because cooler gases give off less energy. It is known that the temperature of the photosphere is around 6,000°C (11,000°F): sunspots are around 1000°C (1,800°F) cooler. It has also been known for over a century that the Sun undergoes 11-year cycles of activity. During maximum periods of activity, sunspots proliferate quite noticeably. The next solar maximum will occur in 1991.

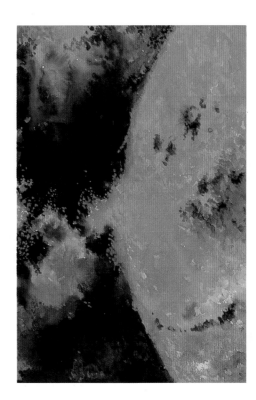

Left: This artwork depicts the dramatic outbreak of sunspots in 1947.

Mercury

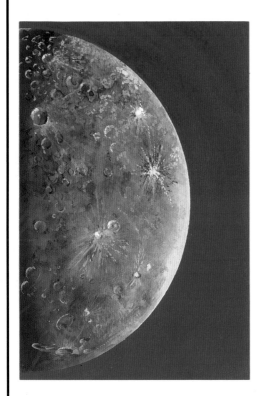

Right: An artist's impression of the scorched surface of Mercury.

Average Distance from Sun: 57.9 million km (35.98 million miles); **Year:** 88 days; **Day:** 25 days; **Diameter:** 4,878km (3,031 miles); **Mass:** 0.06 × Earth; **Surface Gravity:** 0.38 × Earth; **Mean Temperature:** 350°C (660°F) (day): −170°C (−270°F) (night)

The word planet comes from the Greek *planetos* meaning wanderer, because our ancestors believed the planets to be stars that literally wandered through the skies. Unlike the more distant stars, which always appear in roughly the same position, the planets move around because they are so much nearer. Mercury is the nearest planet to the Sun, and is therefore a fairly bright object in the skies. But as it is so near to the Sun, it can only be seen for a few minutes just before dawn or just after dusk. As it is a difficult object to observe, knowledge of the planet remained shrouded in mystery until recently.

In 1974 and 1975, NASA's Mariner 10 made three passes of the planet and revealed its surface to be heavily cratered, making it look vaguely similar to our own Moon. The craters are remnants of the Solar System's birth, when the planets formed out of a vast region of dust and gas which clumped together into planet-sized objects. Material left over from this process has caused cratering throughout the Solar System. Recently, scientists have suggested that Mercury is only half its original size due to a large body impacting it during its formative period.

Mercury has a relatively large, dense core, which produces only a weak magnetic field. Because of its small size, it cannot hold on to much of an atmosphere; what there is is mainly helium.

Venus

Average Distance from Sun: 108 million km (67.11 million miles); **Year:** 225 days; **Day:** −243 days (Venus rotates in the opposite direction to the other planets, indicated here by the minus sign); **Diameter:** 12,102km (7,520 miles); **Mass:** 0.82 × Earth; **Surface Gravity:** 0.9 × Earth; **Mean Temperature:** 480°C (900°F) (surface)

Venus is the second planet from the Sun and is the most brilliant of the planets. It can be seen in the sky after Sunset or before Sunrise, and through

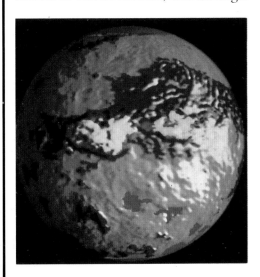

a telescope it exhibits phases like those of the Moon. It is often referred to as our sister planet because it is roughly the same size, though in reality, Venus could not be more different.

It spins in the opposite direction to the other planets, and its day is actually longer than its year – it takes longer to spin on its axis than it does to orbit the Sun. It is enveloped by a dense, broiling carbon dioxide atmosphere which totally covers the surface, trapping heat so effectively that lead would actually melt on its surface. Scientists point to Venus as an example of the greenhouse effect gone wild, when heat cannot escape from the atmosphere and raises the planet's temperature. Space probes have shown clouds of sulphuric acid in which lightning and thunderclaps continuously occur.

The only way its surface can be seen is with radar, a technique which has been successfully performed from spacecraft. Continent-sized plains, highland regions and lowlands have all been seen, with the highest mountains rising some 11km (6.8 miles) above the surrounding terrain, suggesting it is volcanic in origin. Volcanoes similar to the Hawaiian volcanoes are thought to be active.

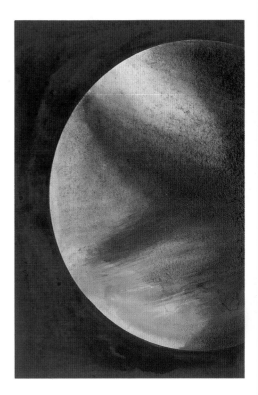

Left: This false colour picture of Venus shows, in green, the highland area, Aphrodite. Lower elevations are in yellow.

Earth

Right: Satellite photograph of the remote regions of the border of India and Tibet.

Average Distance from Sun: 150 million km (93.21 million miles); **Year:** 365.3 days; **Day:** 23 hours 56 minutes; **Diameter:** 12,760km (7,929 miles); **Mass:** 5.98×10^{24}kg; **Surface Gravity:** 1 × Earth; **Mean Temperature:** 20°C (68°F)

The space age has redrawn the portrait of our planet, so the Earth can now be placed in its true planetary perspective. The other inner planets – Mercury, Venus and Mars – are often described as the terrestrial planets, as they are relatively small, rocky bodies like the Earth itself. Space missions have shown that processes which have shaped the Earth have also occurred on them. But the Earth is far more geologically active, as indeed is its atmosphere, thus the surface of the planet has been continuously eroded.

Seventy per cent of the Earth's surface is covered by ocean, which has a fundamental effect on climate and weather. During the 1990s, an international research program will be underway to look at how the oceans and atmosphere interact. Only by monitoring from orbit will it be possible to see if our climate is being altered irreparably.

Unlike the other inner planets, the Earth's core is large and dynamic enough to provide energy for plate tectonics, the relative motion of the upper layers of crust which causes continental drift. The Earth's core is also powerful enough to generate a strong magnetic field which then traps the particles of the solar wind into vast radiation belts, known as the van Allen belts. They effectively protect the Earth's surface from the harshest outbursts from the Sun, though the interaction between the particles and the magnetic field does effect the weather.

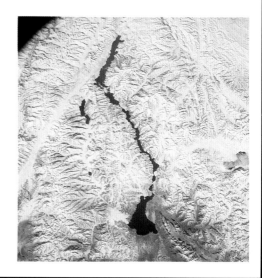

The Moon

Radius of Orbit Around Earth: 385,000km (239,239 miles); **Orbital Period:** 27.3 days; **Diameter:** 3,475km (2,159 miles); **Mass:** 0.0123 × Earth; **Surface Gravity:** 0.16 × Earth

The Moon is the Earth's natural satellite: because its diameter is a quarter that of the Earth, it is quite acceptable to class the Earth and Moon as more of a double planet system. The Moon's phases are well known to even the casual observer of the skies, and they are a consequence of the fact

that the daylight side of the Moon does not always face the Earth. The Moon is spinning about its axis like the Earth, but it completes one rotation in the time it takes to complete one orbit around the Earth. This 'captured' rotation means that only one hemisphere of the Moon is ever seen from Earth. Only since the space age has the Moon's far side been seen in any detail. The Russian Luna 3 obtained the first pictures in 1959.

Unmanned spacecraft and the Apollo manned missions have allowed the Moon to be scrutinized in detail. It has a relatively small core which may extend as far as 1,200km (745 miles) and may also be molten. The uppermost crust is not of uniform thickness being much thicker on the far side, which is also seen to be heavily cratered. Most of the craters on the Moon are the result of impacts from asteroids and meteorites, though some may have been caused by vulcanism. The great 'seas' on the nearside were caused by volcanic activity, though no activity has been seen for millennia. The Moon has a very tenuous atmosphere which could be contained inside a house before its pressure became anything like that on Earth.

Left: The Lunar Roving Vehicle and Apollo 15 seen against the Hadley-Appenine range.

Mars

Right: Viking 2 Lander on the Martian desert at Utopia Planitia in 1976.

Average Distance from Sun: 228 million km (141.68 million miles); **Year:** 687 days; **Day:** 23 hours 37 minutes; **Diameter:** 6,786km (4,217 miles); **Mass:** 0.11 × Earth; **Surface Gravity:** 0.38 × Earth; **Mean Temperature:** −23°C (−9°F)

The next planet out from the Sun after Earth is Mars, named after the Roman god of war because it appears as a red star in the night skies. Through a telescope the planet appears as an orange disc on which darker surface features can be seen. By measuring the time these features take to move around the planet, 17th-century astronomers realized that the Martian day was just under 40 minutes longer than our own. Telescopic observers also observed polar ice caps that advanced and retreated with the seasons, caused by the tilt on Mars on its axis by 23° – virtually the same as the Earth. Dust storms often engulf the whole of the planet for a few months.

Spacecraft in orbit around the planet have discovered that Mars has a range of geological features that would dwarf those on Earth. These include a volcano called *Olympus Mons* that is three times higher than Everest. It is the largest volcano in the Solar System and has been extinct for millions of years. There is a giant canyon nearly 4000km (2,485.5 miles) long named *Valles Marineris* which is 250km (155.3 miles) across in places and as deep as 4km (2.5 miles) with enormous landslides and hanging valleys in its walls.

There is no evidence that life exists on Mars or has ever existed, though there are enormously long, dried-up river channels which were caused by catastrophic flooding billions of years ago, which suggest that once conditions were more Earthlike.

Phobos and Deimos

PHOBOS
Radius of Orbit Around Mars: 9,380km (5,829 miles; **Orbital Period:** 7 hours 40 minutes; **Diameter:** 20 × 23 × 28km (12 × 14 × 17 miles); **Surface Gravity:** 10m per second (32 ft per second)

DEIMOS
Radius of Orbit Around Mars: 23,500km (14,600 miles); **Orbital Period:** 30 hours 26 minutes: **Diameter:** 10 × 12 × 16km (6 × 7 × 10 miles); **Surface Gravity:** 6m per second (20 ft per second)

Mars has two tiny moons called Phobos and Deimos which were discovered in 1877 by an American astronomer named Asaph Hall. They are named after the acolytes of the god of War: Fear and Panic respectively. Our view of them comes from NASA's Viking spacecraft in the late 1970s, and the Soviet Phobos spacecraft which sadly failed in March 1989, before it could return really detailed information.

Both moons are irregularly-shaped objects, with Phobos at most 22km (13.67 miles) in length and Deimos, 12km (7.46 miles). Both are among the darkest objects in the Solar System, reflecting only 6% of the light that falls upon them. Their density is so small that their surface material has been described as more marshmallow than rock.

Both are in near-circular orbits around the Martian equator and are also near to theoretical limits where their orbits become unstable. Phobos is only 5,900km (3,666 miles) from the surface of Mars, and astronomers know its orbit is gradually decaying so that in 70 million years it will eventually spiral inwards towards Mars itself. Both moons are probably captured asteroids.

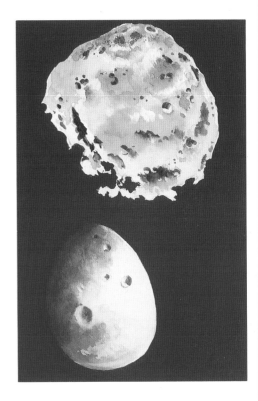

Left: Russian flight control centre for long-term flights monitoring the 'Phobos' probe, July 14, 1988.

The Asteroids

On the very first day of the nineteenth century, an astronomer called Giuseppe Piazzi discovered a small planet between the orbits of Mars and Jupiter. It was the first minor planet or asteroid to be seen and was named Ceres, after the goddess of fertility who was also the patron saint of Sicily from where Piazzi was observing. Over 3,000 asteroids are now known with countless more which cannot be seen: only Vesta, the fourth to be discovered, can be seen with the naked eye. They range in size from Ceres, 100km (62 miles) in diameter, down to a few kilometres.

The combined mass of the known asteroids makes up far less than that of our own Moon, and astronomers believe they are the fragments of a planet that was not able to form. The most likely 'culprit' for this failure is Jupiter, whose immense gravitational pull ensures that the asteroids are herded into distinct groups. Their material is more or less in the same state as when it was formed 4½ billion years ago, and may give clues as to the Solar System's origins. A number of space agencies are hoping to return asteroidal material back to Earth in the 1990s.

Not all the asteroids are contained between Mars and Jupiter: some are found in extremely peculiar orbits. Icarus is an example of an Earth-crossing asteroid, whose orbit takes it closer to the Sun than Mercury. An asteroid called Chiron has been found between the orbits of Saturn and Uranus and little is known about it.

About one per cent of the known asteroids have orbits that cross those of one or more planets. It is estimated that about 1,000 asteroids which cross Earth's orbit are larger than 1km in diameter and will eventually collide with us.

Right: Artist's impression of the asteroid Icarus glowing red as it approaches the Sun.

Jupiter

Average Distance from Sun: 778 million km (483 million miles); **Year:** 11.9 years; **Day:** 9 hours 55 minutes; **Diameter:** 142,800km (88,736 miles); **Mass:** 318 × Earth; **Surface Gravity:** 2.6 × Earth; **Mean Temperature:** −150°C (−240°F) (cloudtops)

The fifth planet from the Sun, Jupiter, is well named after the king of the gods, for it contains nearly 90% of the material of the planets. Unlike the inner, Earthlike planets, Jupiter is a vast ball of hydrogen and helium gas

that rotates in less than 10 hours. Along with Saturn, Uranus and Neptune, it forms a group known as the 'outer planets' or 'gas giants'.

Jupiter has been described as being more like a star that failed than a planet: its interior is hotter than the surface of the Sun. This tremendous heat seeps upwards through the planet's interior and causes Jupiter's atmosphere to move with unbelievable turbulence. Telescopic observers on Earth had noticed long ago that the visible face of Jupiter was constantly changing, with dark regions called 'belts' and lighter 'zones' running parallel to its equator. Four NASA spacecraft which have flown past Jupiter to date (Pioneers 10 and 11, Voyagers 1 and 2) have shown that gas is descending in the belts and ascending upwards in the zones. Vast storm systems have been observed, including the Great Red Spot which is three times larger than the Earth and has been in existence for centuries.

Jupiter also has a vast magnetic field which traps the solar wind into lethal radiation belts, the largest structure seen in the solar system so great is its extent. The Voyagers also discovered a ring around the planet, but this is too faint to be seen from Earth.

Left: Jupiter's recently discovered ring seen here through the upper atmosphere.

IO

Right: A computer-enhanced view of Io where at least eight active volcanoes were found.

Radius of Orbit Around Jupiter: 421,000km (261,609 miles); **Orbital Period:** 1,77 days; **Diameter:** 3,643km (2,264 miles); **Mass:** 1.2 × Earth's Moon

In 1610, the Italian astronomer Galileo Galilei became the first person to focus a telescope on the night skies. As well as seeing the mountains of the Moon and the phases of Venus, he observed four satellites in orbit around Jupiter. They were named the Galilean moons in his honour, and even before the Space Age, they were known to be as large as our own Moon. As telescopes improved, further moons around Jupiter were seen, but they all remained elusive points of light. Mankind's first detailed views came in 1979 thanks to NASA's Voyager spacecraft on their odyssey through the Solar System.

The Voyagers increased the total number of Jupiter's moons to 16, but it was the Galileans that attracted most attention. They were so different that one scientist described them as being a solar system of their own. By far the most puzzling was Io, the innermost of the four, as it was found to be volcanically active. At least eight volcanic plumes were seen spewing sulphurous material high above the surface which continually sprays the surrounding terrain. Io has been described as looking like a strange (and particularly unappetizing) pizza. Its range of colours are caused by the volcanic sulphur.

It is believed that Io is so volcanically active because it is affected by gravitational tides produced by Jupiter and the other Galilean moons. Underneath the crust is a sea of sulphur and sulphur dioxide which is heated by the tidal interactions thus causing eruptions.

Europa

Radius of Orbit Around Jupiter:
671,000km (416,960 miles); **Orbital Period:** 3.56 days; **Diameter:** 3,130km (1,945 miles); **Mass:** 0.66 × Moon

The Galilean moons of Jupiter have been revealed to be far stranger than had ever been imagined: innermost Io is the most volcanically active body ever seen; outermost Ganymede and Callisto are heavily cratered worlds. And Europa, Io's neighbour, has the least cratered surface yet seen – a yellow-coloured icy crust criss-

crossed by shallow grooves which is virtually devoid of craters. It is believed that the crust is a layer of water ice at most 100km (62.14 miles) deep, covering an ocean above a central, rocky core. The tidal effects of Jupiter are probably responsible for cracking the surface which then continually freezes over, although heat from the core may keep cracks open for years.

Taking these facts one stage further, some astronomers suggest that this subterranean sea could contain primitive lifeforms which would receive sunlight when cracks open in the surface ice. The icy crust would effectively act as an insulating blanket to keep the sea warm from the sunlight as well as heat from the radioactive decay of its core. Primitive lifeforms might congregate in 'oases' in the same way that microbes do in lakes in Antarctica on Earth.

Whether this fantastic – and controversial – idea is true or not, Europa is still a curious world in its own right. The next chapter in mankind's exploration of Jupiter will occur in the mid-1990s, when NASA's Galileo spacecraft will spend nearly two years in orbit around the giant planet, making at least one very close pass of each moon in turn.

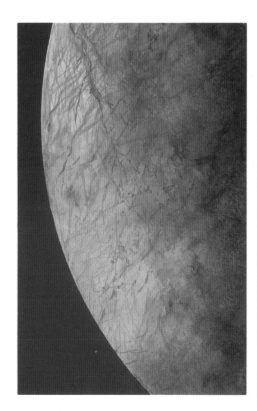

Left: The Galilean moons of Jupiter, clockwise from top, Ganymede, Io, Europa, Callisto.

Saturn

Average Distance from Sun: 1,427 million km (887 million miles); **Year:** 29.46 years; **Day:** 10 hours 40 minutes; **Diameter:** 120,660km (74,978 miles); **Mass:** 95.2 × Earth; **Surface Gravity:** 1.2 × Earth; **Mean Temperature:** 180°C (−290°F) (cloudtops)

The astronomer Brian Marsden once remarked that to the man in the street, the Solar System consists of Mars, Halley's Comet and Saturn's rings. Thanks to man's ingenuity, all have been observed from close range with spacecraft: and it is probably true to say that the last are the most mysterious and spectacular. They were first seen by telescopic observers in the 17th century, who noted that they tilted along with the seasons on Saturn. Three bright rings are easily discernible from Earth, though in the 19th century it was shown mathematically that they could not be solid.

The rings extend nearly 300,000km (186,420 miles) across, though they are only a few kilometres deep. Many thousands of individual rings were seen, each consisting of ice and rock fragments which some astronomers believe are the remnants of a moon which shattered under Saturn's gravitational influence. Within the rings are many gaps and divisions as well as fainter rings which have some very strange properties. One of the outermost rings was found by the Voyagers to be 'kinked' – made up of several individual rings which are intertwined because of the gravitational influence of two small moons.

Saturn itself is mainly made up of hydrogen, with a molten core: its density is so small that if a large enough expanse of water could be found, the ringed planet would actually float!

Right: Saturn and its rings taken from 2.1 million miles by Voyager 2.

Titan

Radius of Orbit Around Saturn:
1,221,850km (759,258 miles); **Orbital Period:** 15.9 days; **Diameter:** 5,150km (3,200 miles); **Mass:** 1.8 × Moon

Saturn's largest moon, Titan, was discovered by Christiaan Huygens in the 17th century, and is by far the most unusual of the Saturnian moons. It is larger than the planet Mercury, and thanks to NASA's Voyager spacecraft, we have detailed information about its structure. Titan has a dense atmosphere with orange clouds

that totally cover its surface.

The Voyagers revealed that up to 95% of Titan's atmosphere is made up of nitrogen, with smaller quantities of methane, ethane, acetylene, propane and even hydrogen cyanide. These complex organic compounds mean that Titan's atmosphere is similar to conditions when life formed on Earth, though with a surface temperature of around −180°C (−290°F) Titan is very much colder. It may be warmed by a slight greenhouse effect. The atmospheric pressure is 1½ times that of our own, which combined with the measured temperature, is known to be near the 'triple point' of methane, where it can exist as either a solid, liquid or gas. So there might be cliffs of methane or seas of methane, or possibly even methane rain or snowflakes beneath the clouds. Titan does not appear to possess a magnetic field and so may not have an electrically conducting core.

Our picture of Titan will be redrawn within the first years of the next century, when a joint NASA/ European Space Agency probe named Huygens will arrive there. Its sensors will take direct measurements of the atmosphere and reveal the true nature of the surface.

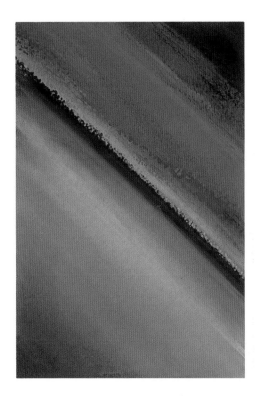

Left: Artist's impression of the surface of Titan and its dense, poisonous atmosphere.

Saturn's Moons

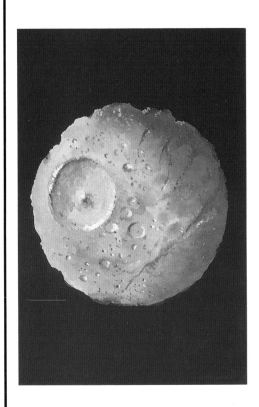

Right: Saturn's Moons. Dione foreground. Tethys, Mimas distant right (artwork above), Enceladus, Rhea (ring's left), Titan.

MIMAS

Radius of Orbit Around Saturn: 186,000km (115,580 miles); **Orbital Period:** 0.94 days; **Diameter:** 398km (247 miles); **Mass:** 0.0005 × Moon

Though the Voyager spacecraft only produced 'snapshots' of Saturn and its moons as they headed out of the Solar System, their information has been revelatory. For the first time, Saturn and its moons have been seen in detail and the tally of moons is now 17. Eight of these are small, irregular bodies, which were swept up at some time by the planet's immense gravitational field.

With the notable exception of Titan, the remainder are icy bodies with heavily cratered surfaces that bear testament to bombardment by asteroids and fragments from the Solar System's birth. By far the most memorable of these icy worlds is Mimas. Only 398km (247 miles) in diameter, Mimas is dominated by a vast crater 130km (80 miles) across – nearly a third the size of the satellite itself. This giant crater has been named after William Herschel, the astronomer who discovered Mimas in the late 18th century. In places it is 10km (6 miles) deep with a central peak 6km (307 miles) high.

Some Voyager photographs have led scientists to jokingly refer to Mimas as being similar to the Death Star from *Star Wars*, but Mimas is a dead, heavily cratered world that has undergone very little change since its formation. The other large Saturnian moons show evidence of internal geological activity, but there is none on Mimas. The object which gouged out the crater Herschel probably disrupted the moon's geological development. In the cold of the outer solar system it is a wonder that Mimas did not shatter.

Uranus

Average Distance from Sun: 2,870 milliom km (1,783 million miles); **Year:** 84 years; **Day:** 17.24 hours; **Diameter:** 52,400km (32,560 miles); **Mass:** 14.5 × Earth; **Surface Gravity:** 0.93 × Earth; **Mean Temperature:** −215°C (−355°F) (cloudtops)

Uranus, the seventh planet from the Sun, was discovered as recently as March 1781 by an amateur astronomer called William Herschel. His discovery literally doubled the size of the solar system overnight, for Uranus is twice as far from the Sun as Saturn.

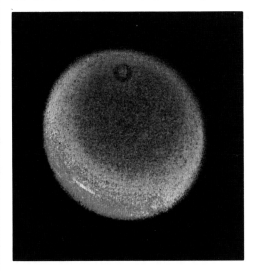

Our current picture of Uranus comes from the Voyager 2 spacecraft which flew past the planet in January 1986, and also took the tally of Uraniam moons to 15 and rings to 11.

The planet itself is one of the strangest of the planets. It is much smaller than Jupiter or Saturn, and seems to have a much smaller internal heat source. It is denser than the larger gas giants, so may contain heavier elements like carbon and iron as well as the hydrogen and helium. It has been said that Uranus rotates on its side, for its axis of its rotation is inclined at 98° to its orbit around the Sun. At the time of the Voyager encounter, the south pole was facing towards the Sun, yet Voyager found that the cloudtops were at the same temperature, −215°C. Because the poles receive much more sunlight than the equatorial regions it had been expected they should be much warmer. Noon on Uranus is the same as dusk on Earth in terms of sunlight.

The magnetic field of Uranus is inclined at 60° to the axis or rotation. Because of the peculiar tilt of the planet itself, the magnetic field is swept into a strange corkscrew-shape that spirals in synchronization with the planet's rotation.

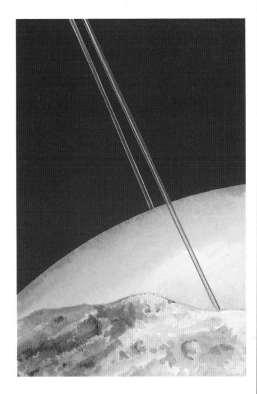

Left: Cloud in the upper atmosphere of Uranus, taken during the Voyager 2 encounter.

Uranus' Moons

MIRANDA
Radius of Orbit Around Uranus: 129,800km (80,658 miles); **Orbital Period:** 1.41 days; **Diameter:** 472km (293 miles); **Mass:** 0.005 × Moon

In the same way that Jupiter and Saturn are surrounded by vast retinues of moons, Uranus has five larger satellites (all of which can be seen with larger telescopes from Earth) and ten smaller fragmented moons. The latter are close to the planet's rings and are probably remnants of material that did not form into rings. The larger moons are named Miranda, Ariel, Umbriel, Titania and Oberon outwards from the planet. They are located in equatorial orbits around Uranus, which, because of its 98° tilt, means they appear in an almost 'bullseye' pattern.

As Voyager 2 headed through the Uranian system, it returned extremely detailed pictures of all of them. Most were icy, displaying evidence for limited geographical activity, but when detailed views of Miranda were returned, they displayed a staggering range of geological features – faults, criss-crossed grooves, and canyons. Cliffs nearly 20km (12.4 miles) high were also seen, presumably made of ice. Quite how such a small body had enough internal energy to create these features remains a puzzle. Some geologists believe that Miranda is the aggregate of a number of moons that were shattered by the gravitational pull of Uranus. Alternatively, tidal interactions with Uranus may be the source of the energy. It is reasonably certain that the activity happened long ago as the surface has been heavily cratered ever since. Oberon, Titania and Umbriel are the least geologically active of the satellites, with Umbriel the darkest and most inactive.

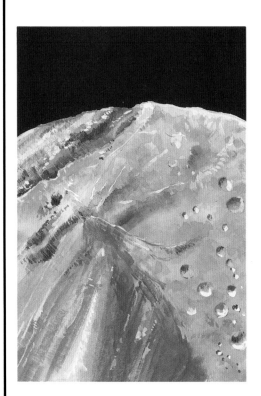

Right: A high-resolution photograph of Uranus' moon, Oberon, taken from Voyager 2.

Neptune

Average Distance from Sun: 4,500 million km (2,800 million miles); **Year:** 165 years; **Day:** 17 hours 52 minutes; **Diameter:** 50,460km (31,356 miles); **Mass:** 17.1 × Earth; **Surface Gravity:** 1.22 × Earth; **Mean Temperature:** −220°C (−364°F)

The eighth planet from the Sun is Neptune, the first planet to be discovered by mathematical calculations. After the orbit of Uranus was worked out, astronomers noted that it deviated slightly from its expected position. Sometimes it appeared as though it was dragging in its orbital motion: other times it appeared to be accelerating. The conclusion was that another planet, further out was the cause. And in 1846 the planet was found exactly where it had been predicted.

But its vast distance from us meant that little was known about it until Voyager 2 reached it in August 1989. Neptune is slightly smaller than Uranus and has a greater density; unlike the bland, featureless Uranus, Neptune's surface is mottled by clouds. Neptune has an internal heat source which "drives" the weather systems, including a dark spot similar to the Great Red Spot on Jupiter. Voyager also observed a higher layer of white, icy clouds which cast shadows on the main cloud deck.

Perhaps the greatest surprise from Voyager was that Neptune has three complete rings. Because its largest moon, Triton, orbits the planet in the opposite direction to Neptune's rotation, the gravitational influence had been expected to disrupt the rings. However, the density of the material within the rings is varied and distinctly clumpy in certain areas. Like the rings of Uranus, they are very dark.

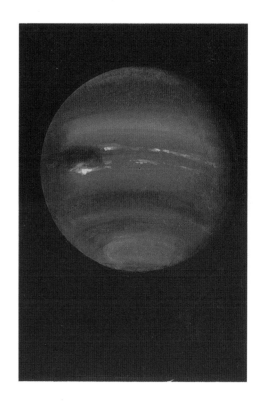

Left: This false-colour image of Neptune taken from Voyager 2, reveal a widespread haze that covers the planet.

Neptune's Moons

TRITON
Radius of Orbit Around Neptune: 354,300km (220,162 miles); **Orbital Period:** 5.88 days; **Diameter:** 3,600-6,000km (2,237 – 3,728 miles); **Mass:** 0.8 × Moon

Before Voyager 2's arrival at Neptune, only two moons had been seen in orbit around the planet. Voyager added a further six moons to the astronomers' inventory, though most were undistinguished objects. Interestingly, one of these new moons is larger than Nereid, the second moon to have been discovered. Nereid, too, was found to be a dark and undistinguished body.

Triton, Neptune's largest moon, was an entirely different matter. Earth-based observations had revealed it to be fairly large and probably surrounded by an appreciable atmosphere. Ironically, Voyager showed otherwise: Triton is smaller than our own Moon and its atmosphere is at least one hundred thousandth the surface pressure on Earth. However, the atmosphere is mainly composed of nitrogen, which is unusual.

It was the moon's surface, however, which stole the show of Voyager 2's encounter. Because it is so cold, the surface is icy making it appear bright with certain regions scarred by grooved terrain. At its south pole, methane has frozen onto the surface giving it a delightful pinkish tinge. After the Voyager encounter, computer analysis of some pictures revealed the presence of a 'plume' of material emanating from what appeared to be a volcanic vent. Though not as immediate as the volcanoes on Jupiter's Io, the geyser-like volcanism on Triton was equally as surprising and suggests an internal energy source.

Right: Two dozen individual images were combined to produce this view of the Neptune-facing hemisphere of Triton. (Voyager 2).

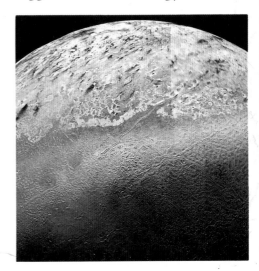

Pluto and Charon

PLUTO
Average Distance from Sun: 5,900 million km (3,666 million miles); **Year:** 248.6 years; **Day:** 6.4 days; **Diameter:** 2,200km (1,367 miles); **Mass:** 0.00017 × Earth?; **Surface Gravity:** 0.03 × Earth?

CHARON
Radius of Orbit Around Pluto: 19,700km (12,242 miles); **Orbital Period:** 6.4 days; **Diameter:** 1,212km (753 miles); **Mass:** 0.02 × Moon?

From analyses of Neptune's orbit it became clear that it too was being accelerated and decelerated in its orbit around the Sun, which led to a further search for another planet. It was found in January 1930 by an assistant at the Flagstaff Observatory in Arizona named Clyde Tombaugh. He became the first, and to date only, person this century to discover a planet.

It soon became clear, however, that Pluto was not a gas giant but rather a small, icy world. Its orbit was also seen to be peculiar: unlike most of the other planets which orbit in the same plane as the Sun, Pluto's orbit is inclined at 17° to that plane, is irregular and comes within the orbit of Neptune as it has been since 1979. So Neptune will actually be the outermost planet until 1999.

In 1978, a moon of Pluto was detected which was later named Charon. By monitoring their motion, their densities have been estimated and both seem to be made up of ice with methane also being detected on the surface of Pluto. But being in the outer recesses of the Solar System, where the Sun is but a reasonably bright star, it is very difficult to obtain information on their masses.

Left: An artist's impression of Pluto (foreground) and Charon at the edge of the Solar System.

Planet X

The very smallness of Pluto and Charon cannot account for the observed perturbations in the orbits of Uranus and Neptune. So since the 1930s, astronomers have been searching for evidence for a tenth planet, usually known by its roman numeral. Historical observations have shown that the perturbations were more pronounced a century ago when Planet X may have been nearer Neptune. Galileo actually observed Neptune in 1613 with a telescope but did not recognize it as another planet.

The search for Planet X has also been aided by NASA's Pioneer 10 and 11, launched a year apart in the early 1970s as pathfinders for the Voyagers in their exploration of the outer Solar System. Both craft are travelling faster than 40,000km/h (25,000 mph). This is the velocity needed to escape from the Sun's gravitational influence, and so they will be the first probes to leave the Solar System. By monitoring their trajectory and the way they spin (both craft spin to stabilize themselves), it was concluded in mid-1987 that Pioneer 10 was not being influenced by a tenth planet beyond Pluto.

However, that is not the end of the story. Pioneer 10 has essentially proved that if Planet X exists, it must be in a highly inclined orbit that is virtually at right angles to the plane of the Solar System which would have no measured effect on its motion. Calculations indicate that a tenth planet would be at least 12 billion km (7.5 billion miles) from the Sun and would take at least 700 years to complete one orbit. It would be at least five times the mass of the Earth. Perhaps further analysis on the motion of Pioneer 11 as well as Voyagers will shed further light on this intriguing puzzle.

Right: An artist's impression of the still-to-be-discovered Planet X. **Above:** Pioneer 10.

Halley's Comet

Time Taken to Orbit Sun: 76 years;
Last Appearance: 1986; **Diameter:**
15 × 8 × 8km (9 × 5 × 5 miles);
Mass: 50–100,000 million tonnes

In days of old, the appearance of a comet was taken to be the portent of evil and a sure sign that plague was about to descend upon the Earth. As a result, comets became synonymous with disease and illness. Today we know that this is nothing more than superstition, as comets are little more than icy bodies just a few kilometres across. In 1986, perhaps the most famous comet returned to the inner part of the Solar System – Halley's Comet. It is named after the scientist who realized that it returned to the inner Solar System every 76 years. Edmond Halley looked through the historical records and predicted that it would appear again after his death.

Like the planets, comets orbit the Sun but on extremely elongated orbits. As they approach the Sun the dust and ice at their cores (technically known as the 'nucleus') is heated and given off as a long tail which trails behind it. The tail looks spectacular in the night sky, and virtually always points away from the Sun.

Some comets appear only once in several thousand years, so vast are their orbits. But others are periodic, and of this category, Halley is the brightest and its orbit the most predictable. It was for this reason that five spaceprobes were launched towards it on its return in 1986. They showed that the nucleus is very dark, and only reflects 2-4% of the light from the Sun. It is made up mostly of water ice (84%) with traces of formaldehyde and carbon dioxide. The comet is now on its way out of the Solar System and is destined to return again in 2061.

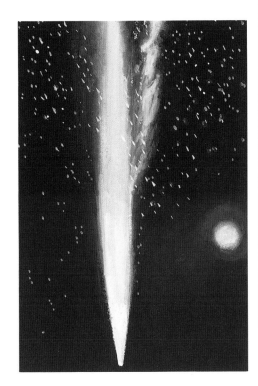

Left: An artist's impression of the spaceprobe Giotto approaching Halley's Comet.

Alpha Centauri

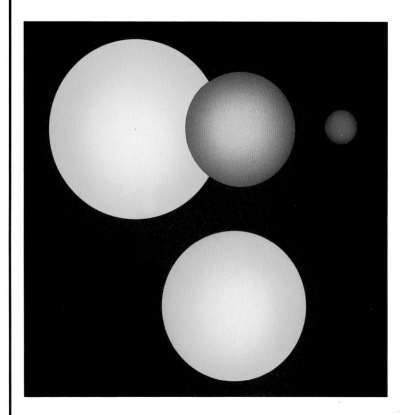

The stars of Alpha Centauri system compared to our Sun (**bottom**). From **left**: Alpha, Beta, Proxima Centauri.

The stars are unimaginably far from Earth – even the nearest one is 402 million million km (250 million million miles) distant. Known as Proxima Centauri, it is part of a grouping of three stars generally known as the Alpha Centauri system. Because distances in space are so vast, astronomers use a unit called the light year to measure how far objects are away from us. So, for example, Proxima Centauri is actually 4.3 light years distant – that is, it takes light from the star 4.3 years to reach us. If you consider that light travels 299,330km (186,000 miles) *per second*, you begin to appreciate the scale of the Universe. Stars in general are very sparsely distributed in space: the average distance between stars is about five light years. Within 17 light years of Earth only 45 stars have been found.

The stars within the Alpha Centauri system are different from each other, each having different ages and properties. Alpha Centauri A – the brightest of the three and the third brightest star in the sky – generates three times as much light as its companions. And though it is a yellow star like our Sun, it is much more massive and luminous. A fainter, orange-coloured star known as Alpha Centauri B orbits the main star and takes 80 years to do so.

Proxima – sometimes called Alpha Centauri C – is a small, fainter star only twice the size of Jupiter. Our Sun emits more light in a week than Proxima does over three centuries, though Proxima does undergo bursts of energy known as flares. Alpha Centauri cannot be seen north of 30° latitude, so is essentially an object for the southern hemisphere observer.

Barnard's Star

The stars move across the night sky for two reasons. Firstly, because the Earth is rotating, and so, like the Sun, they rise in the East and set in the West. And secondly, because they are actually moving in space themselves, though they are so far away that these motions are very tiny. Astronomers call them 'proper motions'. The star with the largest proper motion is called Barnard's Star, the next closest star after the three stars in Alpha Centauri. Because of its relative close-

ness – though still six light years distant! – it moves the diameter of the Moon across the sky in 200 years; quite large, astronomically speaking.

Astronomers class Barnard's star as a red dwarf, which is a small star which contains only 14% of the mass of the Sun. It is far too faint to be seen with the naked eye and requires a telescope to be observed. It was discovered by an American called E. E. Barnard in 1894 and would otherwise be unremarkable apart from an unusual series of observations. Barnard's Star appears to 'wobble' in its proper motion through the sky, but by an amount so small that it requires a large telescope to measure it. These deviations of the motion of Barnard's Star can be explained in terms of two planets orbiting the star in circular orbits with periods of 11.5 and 20-25 years. The innermost of the planets has a mass very close to that of Jupiter, while the outermost has a mass nearly half that of Jupiter.

In June 1989, the satellite HIPPARCOS was launched by the European Space Agency. During the course of its 3-year mission the satellite will measure the proper motions of some 100,000 stars with an accuracy not achieved before.

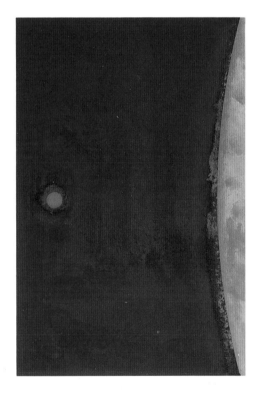

Left: Artist's impression of the System of Barnard's Star.

Ursa Major

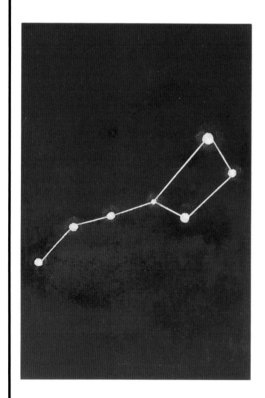

Right: The seven stars forming The Plough in Ursa Major include the naked-eye double seen at the centre of the Plough handle.

A quick glance at the night sky on any clear night shows there are many hundreds of stars visible to the naked eye. Our ancestors studied them intensely, and were quick to recognize groups of stars that seemed to resemble certain shapes, such as animals or mythological figures. Many of these 'constellations' are still in use today, and there is now a grand total of 88 covering the whole of the sky. Many are familiar and obvious, while others are less so.

Perhaps the most well known is Ursa Major or 'The Great Bear', which is nearly always visible in the northern hemisphere. Part of this constellation is familiar as a group of seven stars known as The Plough (in Britain) or The Big Dipper (in the U.S.). It is a convenient 'signpost' to the other stars: for example, extending an imaginary line from the two end stars of the Plough (sometimes known as 'The Pointers') leads to the north pole star, Polaris.

Over the space of a few hours the position of the Plough changes: it appears to move in a circle around the north pole star. This apparent movement is caused by the Earth's rotation on its axis. In fact, Polaris acts almost like a pivot-point around which all the stars visible in the northern hemisphere seem to move. (Southern hemisphere observers will see the stars move around another pivot-point, not obviously connected with any particular star, but which is near the Southern Cross.) Stars near to the horizon will seem to rise in the East and set in the West (like the Sun), though they too are still 'pivoted' around Polaris.

Sirius

There are about 6,000 stars visible with the naked eye: countless more become visible by use of telescopes. The brightest star in the sky is Sirius, the leading star in the constellation of Canis Major (*The Larger Dog*), and it is therefore sometimes called the Dog Star itself. Sirius is the sixth nearest star to the Earth and is best seen during the wintertime, though from northern latitudes it always appears close to the horizon.

The brightness of a star is termed its 'magnitude', a scale which puts the brightest star first and the dimmest last. However, astronomers know that the stars have differing luminosities, the amounts of light that are actually emitted from them, Sirius is only 26 times more luminous than the Sun: its position at the top of the brightness league table is because it is so near us.

By monitoring the proper motion of Sirius, astronomers became aware that it had an unseen companion. It was not until 1862 that Sirius B was actually seen with the finest telescope of the day, and its mass was computed very quickly. It is about the same mass as the Sun, but is contained within a volume only a few kilometres across. So its density is enormous (one sixth of a tonne per cubic centimetre) and the star is known as a White Dwarf, which is a star that has collapsed inwards in the final stages of its evolution. Sirius B was the first White Dwarf to be discovered.

It should be pointed out that nearly 43% of known stars are double, and unlike Sirius, both components can usually be seen fairly easily in a telescope.

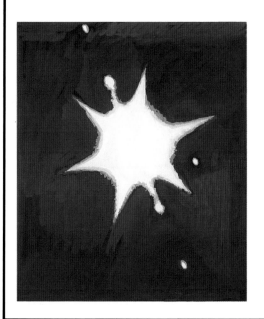

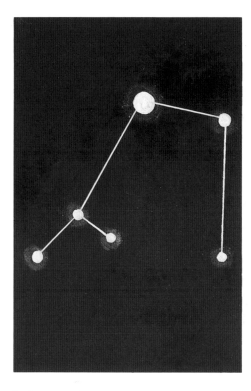

Left: Artwork produced from a photograph taken in 1964 showing Sirius B as a faint image just above Sirius A.

Orion

Because the Earth moves around the Sun, the constellations in the night sky vary from season to season. Some constellations, close to the celestial pole, are visible throughout the year: Ursa Major is the best possible example of such *circumpolar* constellations. Other constellations are only visible at certain times of the year: one of the more obvious during the northern hemisphere winter is Orion, the mighty hunter of mythology. Its distinct line of three stars, representing his belt, can hardly be ignored in the winter skies.

Just below Orion's belt is a region which looks like a patch of bright mist, known as a nebula or 'cloud'. It is a vast expanse of dust and gas, which is sometimes referred to as Orion's Sword. Observations in the infra-red show that it extends much further than is immediately obvious in visible light. Such nebulae are regions in which stars are being born. They are vast stellar kindergartens in which the dust and gas condenses under its own gravity to form objects called proto-stars. But that very dust hides the process of star birth from view, and until recently was one of the great mysteries of astronomy.

Dust particles do not absorb the infra-red radiation, thereby allowing infra-red telescopes to see into the very heart of the nebula. From the way in which they absorb heat, over 50 different molecules have been identified including ice, silicon and even organic compounds such as formaldehyde. Most of the dust is made of carbon in the form of small dust grains, essentially cosmic soot. Other molecules have been revealed by radio telescopes which can 'tune into' their atomic vibrations.

Right: False-colour image of the area around the nebula in Orion.

Beta Pictoris

Star births do not take place overnight: it may take many tens of thousands of years for a proto-star to contract sufficiently for its core to warm up. The more massive the proto-star, the hotter it will become in later life. Stars of greater mass will eventually 'go nuclear' when their core temperature reaches ten million degrees Celsius (18 million degrees Fahrenheit). When this happens, vast streams of radiation are given out which can disrupt the formation of

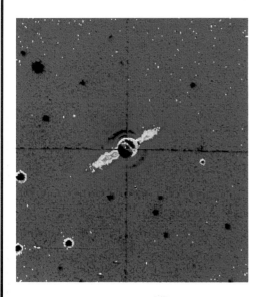

stars in the vicinity. After this violent period of birth, the star will eventually settle down to a stable existence. Protostars are observed at infra-red wavelengths and are found in regions of the Galaxy where there is an abundance of gas and dust.

Yet the star will still be surrounded by an extensive expanse of dust and gas which may clump together eventually to form planets like those in our Solar System. This seems to be happening in the star known as Beta Pictoris, 80 light years away from us. In the infra-red it is emitting 100 times more energy than expected, and this is believed to be caused by dust and debris extending for 80 billion km (50 billion miles) out beyond the star itself.

Optical observations have shown that there is indeed a disc of material which we see edge-on, contained within which are 'depletion' regions. These have been likened to a hole in a record, suggesting that planets are forming which have swept up the dust and gas there. Most of this is believed to be made up of ices, silicates and carbonaceous materials, which happen to be the very materials out of which our own planetary system formed.

Left: Computer-enhanced image of the disc of gas and dust surrounding Beta Pictoris.

Eta Carinae

Stars are vast balls of hydrogen and helium which shine by the same nuclear transformation which powers hydrogen bombs on Earth. When one gram of hydrogen is converted to helium the energy released is equivalent to that produced in over 35 billion large, hydroelectric power stations in one year. As something like 90% of the known universe is made up of hydrogen, perhaps we should not be too surprised to find so many stars.

Astronomers know that the more massive a star is, the greater the temperature and pressure at its core and the faster the rate at which hydrogen is converted to helium. This means more energy is given off and consequently the star is hotter and brighter. In fact, the life of a star is more or less determined by how heavy it is when it is formed. A heavy star converts the hydrogen into helium so quickly that it soon exhausts its supply. A lighter star burns the hydrogen less prodigiously and so lasts longer. Our sun is midway between these two extremes.

Astronomers believe that the heaviest star in the sky is Eta Carinae, which can be observed only from southern latitudes. It is believed to be at least 8,000 light years away and have a mass of one hundred Suns. Its luminosity is believed to be equivalent to at least 6 million Suns, and during the mid-19th century it was the second brightest star in the sky. Today it is too faint to be seen with the naked eye, and through a telescope appears red, surrounded by a white nebulous patch. In the infra-red this region is very bright, suggesting that stars are being born there.

Right: The Eta Carinae Nebula located in the Carina spiral arm of the Galaxy is some 400 light years across.

Algol

Many stars are variable – that is they brighten and fade over relatively short periods of time. Many vary in brightness because of subtle changes within their cores, others vary because they are double stars known as 'eclipsing binaries'. This latter category occurs when a larger, fainter star passes in front of a brighter star, and so the total amount of light observed appears to drop. From Earth this is seen as a drastic reduction in the star's brightness. When the fainter star is eclipsed by the brighter, the light reduction is much less.

The most famous eclipsing binary is Algol in the constellation of Perseus. Its variability was recognized in ancient times and its name is Arabic for 'Winking Demon'. Every 2½ days Algol appears to 'wink', and its brightness drops from 2nd magnitude to below 3rd. Astronomers know that the exact period of winking is affected by a third star in the Algol system.

Of the truly inherently variable stars, there are many categories and sub-divisions. Many are named after certain stars which show characteristic behaviours: for example, Mira variables are named after Mira Ceti, for which the brightness variations last between 80 and 100 days.

Bright star

Faint star

Faint star blocks off bright light

Bright light can be seen again

Giants and Dwarfs

Right: Artist's impression of the surface of a planet of a Red Dwarf.

Stars are born, they develop and then they die – but on timescales far longer than for human lives. What happens when stars are coming to the end of their evolution? Astronomers know that heavier stars consume their hydrogen fuel relatively quickly, and have shorter lives and certainly spectacular deaths. For stars which have roughly the same initial mass as the Sun (from 0.1 to 1.4 solar mass), the beginning of the end is when it expands to a hundred times its size and becomes a red giant. One of the most famous examples of red giants is Betelgeuse, seen as Orion's shoulder (top left from northern latitudes). It has a diameter equivalent to 800 Suns and has a mass twenty times that of our star.

Sunlike stars expand to become red giants when all the hydrogen has been used up in their core: as a result, the core shrinks and heats up. Different nuclear reactions occur there with the remaining helium so that the star's outermost layers expand and cool. But red giants cannot hold on to the outer regions for long. Instead they become unstable and the upper regions of their atmosphere disperse into outer space.

Only a tiny core remains, which is no longer producing heat and so cools and fades. These objects are known as White Dwarfs, being 1/100th the diameter of the Sun. The material is packed so tightly that its average density is millions of times that of water. The densities of these stars makes them fundamentally different in nature from 'normal' stars. In our Sun it is the thermal pressures generated that prevent it from collapsing. In White dwarfs it is more fundamental quantum mechanical effects that maintain the star. White dwarfs are a possible evolutionary endpoint.

Supernovae

SN 1987A
Distance from Sun: 170,000 light years; **Mass:** 20 × Sun; **Surface Temperature:** 20,000°C (36,000°F); **Age:** 20 million years

Stars heavier than 1.4 times the mass of the Sun have a much more dramatic end than their lighter brethren: they end their lives with a vast explosion known as a supernova outburst which erupts with as much energy as that generated by whole galaxies.

Once the hydrogen inside the core of a heavier star is used up, the

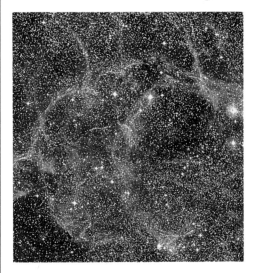

remaining helium fuses into carbon because of the intense pressures and temperatures. This transformation produces even greater amounts of energy, increasing temperatures and pressures and thereby producing further heavier elements. Once this happens, the star's core can best be described as 'onion-like', with distinct, concentric layers of iron, silicon, carbon, helium and the last vestiges of hydrogen running outwards from the core.

When iron is produced, however, energy is needed to form it: energy is not produced, but rather used up, and so the core becomes unstable. In the period of a few seconds, it collapses in on itself and a vast shock wave of energy is produced. This is the supernova outburst.

Around 20 supernova outbursts are detected per year – usually in distant galaxies by dedicated, eagle-eyed amateur astronomers. In recent years, however, a remarkable supernova outburst took place in the Large Magellanic Cloud (see page 42). It was the first supernova visible to the naked eye in nearly four centuries, and became apparent on 23 February 1987 though it was quite puny by supernova standards.

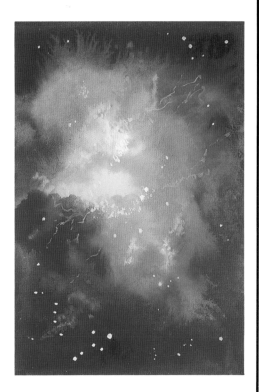

Left: The filamentary structure of the Vela supernova remnant, which originated 12,000 years ago. It has a pulsar at its centre.

Pulsars

During a supernova explosion, most of the swollen star's material is blown into space. All that is left is a core so utterly small that the remaining material is crushed even more densely than that inside a white dwarf. These remnants are referred to as neutron stars, and are at most, about 20km (12.5 miles) across. Even the protons and electrons (positively and negatively charged particles) within them become fused together to form particles without electrical charge (neutrons). The material is so dense that even a pinhead of it would weigh around a million tonnes!

The first neutron star to be discovered was detected by radio means, revealing their very strange properties. In late 1967, radio astronomers at Cambridge University observed a weak radio source that appeared to be 'ticking', so regular were its radio pulses. The Cambridge astronomers realized that the object was spinning rapidly, like a celestial lighthouse in space. Rather than sending out beams of light that seem to flash, these rapidly spinning objects sent out radio pulses. They were called pulsars as a result, and this is the name by which most neutron stars are now known. To account for the number of pulsars in our Galaxy, a neutron star would have to be formed approximately every 50-100 years.

Since then, 300 or so pulsars have been discovered. Most of them were found to be spinning once every second – but in 1982, the first millisecond pulsar was discovered. Known as PSR 1937 + 215, it rotates every 1.557 milliseconds. Further millisecond pulsars have been discovered which remain among the more puzzling of mysteries for astronomers.

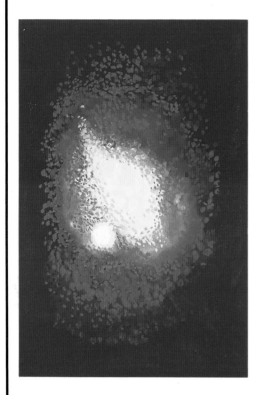

Right: An artist's visualization of the 'lighthouse' theory of pulsar activity.

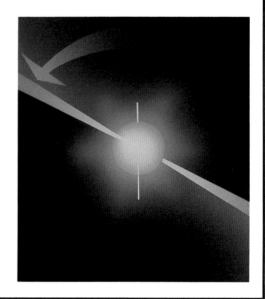

Black Holes

Even more peculiar than pulsars are black holes, objects which astronomers believe do exist though they cannot be observed directly. If a very heavy star (three times the mass of the Sun) explodes as a supernova, its core condenses so rapidly that it implodes under its own gravitation. It is so small and dense that material cannot escape out of it – even light – so it will appear black.

Theoretically a black hole shrinks to nothing and has an infinite density.

However, it will be surrounded by a region a few kilometres across out of which nothing can escape, its extent depending on the original size of the star which has collapsed. Astronomers have calculated that for the Sun it would be around 3km (1.86 miles) across. This 'event horizon', as it is known, defines the limit of the black hole – and the extent of our knowledge, for all laws of physics break down once entered.

Black holes cannot be detected directly, but rather by their interaction with other stars. One of the brightest objects in the sky at X-ray wavelengths is known as Cygnus X-1 after the constallation of Cygnus in which it is found. With optical telescopes, the star which corresponds to this immense source of radiation appears as a relatively harmless object. But it is in orbit around an unseen companion which has a gravitational pull of at least ten Suns. It is believed to be a black hole into which matter from the nearby star is accelerating, causing it to heat and give off streams of X-rays as a result. Black holes may also have formed during the Big Bang when regions of gas and dust became so compressed that they formed so-called primordial black holes.

Left: Artist's impression of the Black Hole believed to be in the constellation of Cygnus.

Star Clusters

Right: The globular cluster Omega Centauri, which is the finest example of such a cluster in the heavens.

Stars are often grouped in association known as clusters, which can be classed into two basic groups. Open or loose clusters have no definite shape and contain at most several hundred stars. Globular clusters are symmetrical and contain up to a million stars. A number of clusters are visible with the naked eye, perhaps the most famous being the Pleiades (Seven Sisters) in the constellation of Taurus.

The Pleiades are a fine example of an open cluster. Through even a small telescope they are surrounded by a distinctly blue tinge. This nebulosity 'cloudiness' is explained by light from the stars being reflected in dust believed to be the mortal remains of the nebula out of which the stars formed around 50 million years ago. In fact most open clusters contain stars which are of roughly the same age, and more often than not hot, white stars like those in the Pleiades.

Most of the hundred or so known globular clusters are located in the southern hemisphere skies. By plotting their positions, astronomers have realized that they are centred on the constellation of Sagittarius, where the centre of our Galaxy is located. They form a 'halo' around our galaxy's extremities and are extremely remote, being at least 20,000 light years away.

The 140-odd known globular clusters belonging to our own Galaxy are extremely old, with ages greater than 10,000 million years. A few globular clusters have been found to contain X-ray sources which may be Black Holes. The number of known open clusters in the Galaxy is 1200, but this is probably only a fraction of the total number. All stars are thought to be born in clusters.

The Milky Way

The Milky Way is the more popular name given to our own Galaxy, because it is seen as a luminous band in the night skies. It is seen to best advantage in the southern hemisphere, though during the northern summer it crosses from one horizon to the other. Through a small telescope or even binoculars, it is seen to be made up of countless stars, dark patches and clusters and nebulae. The fact that it so obviously straddles our skies shows that we live within it.

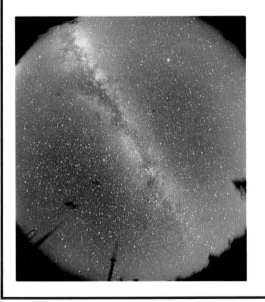

Astronomers estimate that our Galaxy numbers 100,000 million stars, forming a loose spiral or Catherine wheel shape overall. It is a flattened disc which has been best described as two fried eggs back-to-back. The central bulge is believed to be 20,000 light years thick and as a whole may extend for 100,000 light years. The Galaxy is rotating, and it has been calculated that the Sun is moving at a speed of 250km (155 miles) per second. As a result, it takes over 200 million years to complete one rotation. Quite recently, radio observations made at the wavelength of carbon monoxide, and taken in Australia, have painted a clearer picture of our Galaxy which suggests that there are four arms to the spiral. But our view from within the disk can never be wholly adequate.

Our Sun is just one in this multitude of stars, and is about a third of the way out from the Galaxy's centre. That centre itself is in the constellation of Sagittarius, where the Milky Way is seen at its widest, and is obscured by dust. Radio astronomers have been able to look right into this region, where it is believed the utter concentration of material may have led to the formation of a powerful black hole.

Left: This photograph, made with an extremely wide angle lens, shows the Milky Way from horizon to horizon.

The Magellanic Clouds

The Major (top) and Minor Magellanic Clouds.

Right: The Large Cloud of Magellan, a major system in its own right, about a quarter the size of our own Galaxy.

In 1520, the explorer Ferdinand Magellan and his crew became aware of two bright patches of light in the night skies, becoming ever more resplendent as they sailed down the coast of South America. As a result, they are now known as the Large and Small Clouds of Magellan, and are classed as irregular galaxies as they have indeterminate shapes. They are among the brightest nebulous objects in the sky and can still be seen in moonlight.

In earlier times they were thought to be part of our own Galaxy, but we now know they are 'satellite' galaxies, well over 100,000 light years distant. How their distance was measured was of crucial importance in astronomy. In the early part of this century, an astronomer called Henrietta Leavitt at Harvard University realized that a group of variable stars known as the Cepheids can be used as distance indicators. They are so named after the star Delta Cephei which changes in brightness like clockwork over 5.4 days: a period which is related to its actual luminosity. Henrietta Leavitt looked for Cepheids in the Magellanic Clouds to measure their periods of variability. This told her their real luminosities, and since she knew their apparent brightness from Earth, it was a simple calculation to work out how far away they really were. In fact, the Cepheid method has been used to measure the distance of galaxies local to ours.

The most recent estimates indicate that the Large Magellanic Cloud is 160,000 light years away and the Small Magellanic Cloud is 185,000 light years. Astronomers have paid a great deal of attention to them, especially after the supernova outburst in 1987.

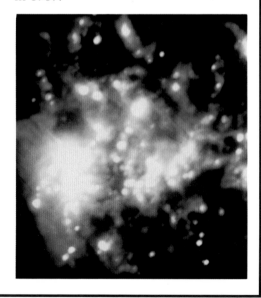

The Andromeda Galaxy

In the same way that stars tend to clump together because of their mutual gravitational attraction, the same occurs with galaxies. Astronomers now know that in our immediate vicinity, galactically speaking, there are around two dozen galaxies which form our Local Group. The Magellanic Clouds are members, but by far the largest is the Andromeda Galaxy, which is some 2.2 million light years distant and 1½ times larger than our own galaxy.

Astronomers have studied it in

detail and have learned much more about how stars and galaxies form. On fine winter evenings in the northern hemisphere it can just about be seen with the naked eye as a faint patch of light. It is sometimes called the Great Galaxy of Andromeda and its spiral shape only becomes evident in larger telescopes.

Details of the spiral structure are hard to determine, however, because the galaxy is so close to edge-on to us. One theory proposes that there are two trailing spiral arms with some disturbance to the nearby galaxy M32. Another theory suggests a single arm set up via gravitational 'resonance' with M32.

By use of the Cepheid method, it was realized that the Andromeda Galaxy was a companion galaxy to our own, though the exact distance had to be redefined in the early 1950s when it became apparent that there were two sorts of Cepheids, and those in the great Galaxy had been misidentified. Its spiral shape is believed to be similar to that of our own, and more detailed studies have revealed a halo of globular clusters three times more extensive than those in our own. Radio observations have shown that there is much less activity at its centre.

Left: Above is an optical view of the Andromeda Galaxy accompanied by a radio map.

Galaxies

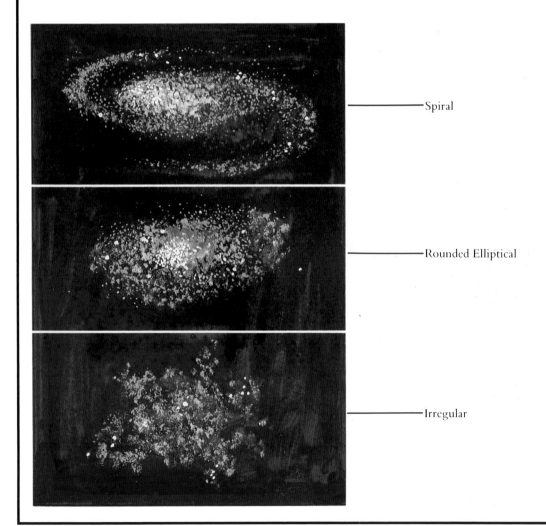

Spiral

Rounded Elliptical

Irregular

Our Galaxy is one of millions and by no means unusual or unique. Astronomers have classified galaxies by their various shapes which include spirals, barred spirals, ellipticals and irregulars. Within each grouping there are subdivisions, but the above descriptions proposed by the American astronomer Edwin P. Hubble in 1925 are reasonably watertight. Our own galaxy is a relatively loose spiral galaxy, similar to Andromeda.

Earlier this century it was realized that galaxies were located far beyond our own, but of far greater significance was the fact that all observed galaxies are seen moving away from us, showing that the whole universe is expanding. This became apparent when their spectra were obtained. By passing light from galaxies through a prism, its component colours were seen as a spread similar to that seen in a rainbow. Lines in the spectra (caused by the elements which they contain) were seen to be shifted towards the red end of the spectrum. This increase in the wavelength was caused by their rapid motion away from the Earth. The faster the galaxies are moving, the farther away from us they are, and some have been observed receding from us at 90% of the speed of light.

Superclusters

In recent years, astronomers have discovered that local groups of galaxies form part of even larger groupings commonly referred to as superclusters. These should not be confused with clusters of stars which are obviously much smaller. That the universe is 'clumpy' was an idea proposed in the 19th century, but evidence for it has only been forthcoming since the late 1950s. Astronomers have mapped the positions and concentrations of galaxies and realized that there are vast clouds of galaxies towards the constellations of Ursa Major and Virgo. These are part of our own supercluster, which stretches for many millions of light years and is made up of many thousands of galaxies.

The concentration of galaxies within superclusters is also clumpy. Towards the constellations of Virgo, astronomers have noted a richness of galaxies numbering a thousand members. It was later realized this Virgo Association was the centre of our own supercluster, which we now know influences the way in which the universe is expanding in this region. Our local group is moving away from the Virgo association, but its motion is literally being held back because of its immense gravitational pull, even though it is 60 million light years away. It is estimated that The Virgo Association has a mass of some 800 billion Suns and stretches nearly 150,000 light years across.

Another association within another supercluster is that in the constellation Coma which is at least 450 billion light years away. It is over 10 million light years across and contains many more galaxies than our own supercluster.

Clusters of galaxies are landmarks that can be traced out to great distances. Their study will be of use in testing cosmological models as well as teaching us about the formation of galaxies.

Above: The Fornax Cluster of galaxies contains both spiral and elliptical galaxies. The cluster is 55 million light years away.

Quasars

The more distant a galaxy is from us, the more difficult it is to observe in any detail. Some of the most remote objects seen by astronomers are among the most luminous, and are known simply as 'Quasi-Stellar Objects' or 'Quasars'. The first to be discovered was known to be a very powerful radio source and catalogued as the 273rd object in the third list of interesting objects produced by radio astronomers at Cambridge University. In the early 1960s, 3C 273 was identified optically as a faint, bluish star, and its spectrum revealed that it had a very large red shift. This was interpreted as meaning that the object was a vast distance away and accelerating very rapidly: yet, galaxies at similar distances were also known, but they were very much fainter.

Further analysis of optical images of 3C 273 reveal a faint luminous 'jet' of material emanating from the quasar itself. This material is at least 150,000 light years in length, and its infra-red power output is over a thousand times that of the Milky Way. Since then, further quasars have been observed, including one which is travelling at 96% the velocity of light in the most distant reaches of the known universe.

Though they still present many theoretical problems, astronomers believe they can be explained as the cores of active galaxies containing massive black holes which can literally feed on the stars and material at its centre. This will produce enough energy to explain the great luminosities observed at radio wavelengths. Because quasars can be seen at greater distances than any other object and so farther back in time, they provide a way of examining the Universe's youth.

Right: The galaxy NGC 4319 with a quasar arrowed. The quasar is apparently no farther away than the galaxy.

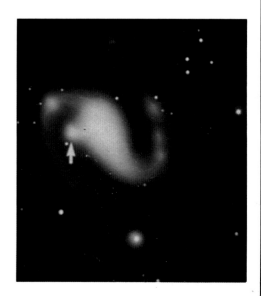